The Compact Org Mode Guide

Release 9.2.4

by Carsten Dominik

Copyright © 2010–2019 Free Software Foundation

Permission is granted to copy, distribute and/or modify this document under the terms of the GNU Free Documentation License, Version 1.3 or any later version published by the Free Software Foundation; with no Invariant Sections, with the Front-Cover Texts being "A GNU Manual," and with the Back-Cover Texts as in (a) below. A copy of the license is included in the section entitled "GNU Free Documentation License" in the full Org manual, which is distributed together with the compact guide.

(a) The FSF's Back-Cover Text is: "You have the freedom to copy and modify this GNU manual."

Short Contents

1 Introduction 1
2 Document Structure 2
3 Tables 6
4 Hyperlinks 8
5 TODO Items 10
6 Tags 13
7 Properties 15
8 Dates and Times 16
9 Capture - Refile - Archive 19
10 Agenda Views 21
11 Markup for rich export 26
12 Exporting 29
13 Publishing 31
14 Working with source code 32
15 Miscellaneous 34
A GNU Free Documentation License 35

1 Introduction

1.1 Preface

Org is a mode for keeping notes, maintaining TODO lists, and doing project planning with a fast and effective plain-text system. It is also an authoring and publishing system, and it supports working with source code for literal programming and reproducible research.

This document is a much compressed derivative of the comprehensive Org-mode manual. It contains all basic features and commands, along with important hints for customization. It is intended for beginners who would shy back from a 200 page manual because of sheer size.

1.2 Installation

Important: If you are using a version of Org that is part of the Emacs distribution, please skip this section and go directly to Section 1.3 [Activation], page 1.

If you have downloaded Org from the Web, either as a distribution `.zip` or `.tar` file, or as a Git archive, it is best to run it directly from the distribution directory. You need to add the `lisp` subdirectories to the Emacs load path. To do this, add the following line to `.emacs`:

```
(setq load-path (cons "~/path/to/orgdir/lisp" load-path))
(setq load-path (cons "~/path/to/orgdir/contrib/lisp" load-path))
```

If you have been using git or a tar ball to get Org, you need to run the following command to generate autoload information. command:

```
make autoloads
```

1.3 Activation

Add the following lines to your `.emacs` file. The last four lines define *global* keys for some commands — please choose suitable keys yourself.

```
;; The following lines are always needed.  Choose your own keys.
(global-set-key "\C-cl" 'org-store-link)
(global-set-key "\C-ca" 'org-agenda)
(global-set-key "\C-cc" 'org-capture)
(global-set-key "\C-cb" 'org-switchb)
```

Files with extension '`.org`' will be put into Org mode automatically.

1.4 Feedback

If you find problems with Org, or if you have questions, remarks, or ideas about it, please mail to the Org mailing list emacs-orgmode@gnu.org. For information on how to submit bug reports, see the main manual.

2 Document Structure

Org is based on Outline mode and provides flexible commands to edit the structure of the document.

2.1 Outlines

Org is implemented on top of Outline mode. Outlines allow a document to be organized in a hierarchical structure, which (at least for me) is the best representation of notes and thoughts. An overview of this structure is achieved by folding (hiding) large parts of the document to show only the general document structure and the parts currently being worked on. Org greatly simplifies the use of outlines by compressing the entire show/hide functionality into a single command, `org-cycle`, which is bound to the `TAB` key.

2.2 Headlines

Headlines define the structure of an outline tree. The headlines in Org start with one or more stars, on the left margin[1]. For example:

```
* Top level headline
** Second level
*** 3rd level
    some text
*** 3rd level
    more text

* Another top level headline
```

Note that a headline named after `org-footnote-section`, which defaults to 'Footnotes', is considered as special. A subtree with this headline will be silently ignored by exporting functions.

Some people find the many stars too noisy and would prefer an outline that has whitespace followed by a single star as headline starters. Section 15.2 [Clean view], page 34, describes a setup to realize this.

2.3 Visibility cycling

Outlines make it possible to hide parts of the text in the buffer. Org uses just two commands, bound to `TAB` and `S-TAB` to change the visibility in the buffer.

TAB
: *Subtree cycling*: Rotate current subtree among the states
```
,-> FOLDED -> CHILDREN -> SUBTREE --.
'-----------------------------------'
```
When called with a prefix argument (`C-u TAB`) or with the shift key, global cycling is invoked.

`S-TAB` and `C-u TAB`
: *Global cycling*: Rotate the entire buffer among the states
```
,-> OVERVIEW -> CONTENTS -> SHOW ALL --.
'--------------------------------------'
```

---

[1] See the variable `org-special-ctrl-a/e` to configure special behavior of `C-a` and `C-e` in headlines.

Chapter 2: Document Structure

`C-u C-u C-u TAB`
    Show all, including drawers.

When Emacs first visits an Org file, the global state is set to OVERVIEW, i.e. only the top level headlines are visible. This can be configured through the variable `org-startup-folded`, or on a per-file basis by adding a startup keyword `overview`, `content`, `showall`, like this:

    `#+STARTUP: content`

2.4 Motion

The following commands jump to other headlines in the buffer.

`C-c C-n`   Next heading.

`C-c C-p`   Previous heading.

`C-c C-f`   Next heading same level.

`C-c C-b`   Previous heading same level.

`C-c C-u`   Backward to higher level heading.

2.5 Structure editing

`M-RET`   Insert new heading with same level as current. If the cursor is in a plain list item, a new item is created (see Section 2.7 [Plain lists], page 4). When this command is used in the middle of a line, the line is split and the rest of the line becomes the new headline[2].

`M-S-RET`   Insert new TODO entry with same level as current heading.

`TAB` in new, empty entry
    In a new entry with no text yet, `TAB` will cycle through reasonable levels.

`M-left/right`
    Promote/demote current heading by one level.

`M-S-left/right`
    Promote/demote the current subtree by one level.

`M-S-up/down`
    Move subtree up/down (swap with previous/next subtree of same level).

`C-c C-w`   Refile entry or region to a different location. See Section 9.2 [Refile and copy], page 20.

`C-x n s/w`   Narrow buffer to current subtree / widen it again

When there is an active region (Transient Mark mode), promotion and demotion work on all headlines in the region.

---

[2] If you do not want the line to be split, customize the variable `org-M-RET-may-split-line`.

Chapter 2: Document Structure

2.6 Sparse trees

An important feature of Org mode is the ability to construct *sparse trees* for selected information in an outline tree, so that the entire document is folded as much as possible, but the selected information is made visible along with the headline structure above it[3]. Just try it out and you will see immediately how it works.

Org mode contains several commands creating such trees, all these commands can be accessed through a dispatcher:

`C-c /` This prompts for an extra key to select a sparse-tree creating command.

`C-c / r` Occur. Prompts for a regexp and shows a sparse tree with all matches. Each match is also highlighted; the highlights disappear by pressing `C-c C-c`.

The other sparse tree commands select headings based on TODO keywords, tags, or properties and will be discussed later in this manual.

2.7 Plain lists

Within an entry of the outline tree, hand-formatted lists can provide additional structure. They also provide a way to create lists of checkboxes (see Section 5.6 [Checkboxes], page 12). Org supports editing such lists, and the HTML exporter (see Chapter 12 [Exporting], page 29) parses and formats them.

Org knows ordered lists, unordered lists, and description lists.

- *Unordered* list items start with '-', '+', or '*' as bullets.
- *Ordered* list items start with '1.' or '1)'.
- *Description* list use ' :: ' to separate the *term* from the description.

Items belonging to the same list must have the same indentation on the first line. An item ends before the next line that is indented like its bullet/number, or less. A list ends when all items are closed, or before two blank lines. An example:

```
** Lord of the Rings
   My favorite scenes are (in this order)
   1. The attack of the Rohirrim
   2. Eowyn's fight with the witch king
      + this was already my favorite scene in the book
      + I really like Miranda Otto.
   Important actors in this film are:
   - **Elijah Wood** :: He plays Frodo
   - **Sean Astin** :: He plays Sam, Frodo's friend.
```

The following commands act on items when the cursor is in the first line of an item (the line with the bullet or number).

`TAB` Items can be folded just like headline levels.

`M-RET` Insert new item at current level. With a prefix argument, force a new heading (see Section 2.5 [Structure editing], page 3).

`M-S-RET` Insert a new item with a checkbox (see Section 5.6 [Checkboxes], page 12).

---

[3] See also the variable `org-show-context-detail` to decide how much context is shown around each match.

Chapter 2: Document Structure

`M-S-up/down`
: Move the item including subitems up/down (swap with previous/next item of same indentation). If the list is ordered, renumbering is automatic.

`M-left/M-right`
: Decrease/increase the indentation of an item, leaving children alone.

`M-S-left/right`
: Decrease/increase the indentation of the item, including subitems.

`C-c C-c`
: If there is a checkbox (see Section 5.6 [Checkboxes], page 12) in the item line, toggle the state of the checkbox. Also verify bullets and indentation consistency in the whole list.

`C-c -`
: Cycle the entire list level through the different itemize/enumerate bullets ('-', '+', '*', '1.', '1)').

2.8 Footnotes

A footnote is defined in a paragraph that is started by a footnote marker in square brackets in column 0, no indentation allowed. The footnote reference is simply the marker in square brackets, inside text. For example:

```
The Org homepage[fn:1] now looks a lot better than it used to.
...
[fn:1] The link is: https://orgmode.org
```

The following commands handle footnotes:

`C-c C-x f`
: The footnote action command. When the cursor is on a footnote reference, jump to the definition. When it is at a definition, jump to the (first) reference. Otherwise, create a new footnote. When this command is called with a prefix argument, a menu of additional options including renumbering is offered.

`C-c C-c`
: Jump between definition and reference.

Further reading
Chapter 2 of the manual
Sacha Chua's tutorial

3 Tables

Org comes with a fast and intuitive table editor. Spreadsheet-like calculations are supported in connection with the Emacs `calc` package (see the Emacs Calculator manual for more information about the Emacs calculator).

Org makes it easy to format tables in plain ASCII. Any line with '|' as the first non-whitespace character is considered part of a table. '|' is also the column separator. A table might look like this:

```
| Name  | Phone | Age |
|-------+-------+-----|
| Peter | 1234  | 17  |
| Anna  | 4321  | 25  |
```

A table is re-aligned automatically each time you press `TAB` or `RET` or `C-c C-c` inside the table. `TAB` also moves to the next field (`RET` to the next row) and creates new table rows at the end of the table or before horizontal lines. The indentation of the table is set by the first line. Any line starting with '|-' is considered as a horizontal separator line and will be expanded on the next re-align to span the whole table width. So, to create the above table, you would only type

```
|Name|Phone|Age|
|-
```

and then press `TAB` to align the table and start filling in fields. Even faster would be to type `|Name|Phone|Age` followed by `C-c RET`.

When typing text into a field, Org treats `DEL`, `Backspace`, and all character keys in a special way, so that inserting and deleting avoids shifting other fields. Also, when typing *immediately after the cursor was moved into a new field with* `TAB`, `S-TAB` *or* `RET`, the field is automatically made blank.

Creation and conversion

`C-c |`   Convert the active region to table. If every line contains at least one TAB character, the function assumes that the material is tab separated. If every line contains a comma, comma-separated values (CSV) are assumed. If not, lines are split at whitespace into fields.

If there is no active region, this command creates an empty Org table. But it's easier just to start typing, like `|Name|Phone|Age C-c RET`.

Re-aligning and field motion

`C-c C-c`   Re-align the table without moving the cursor.

`TAB`   Re-align the table, move to the next field. Creates a new row if necessary.

`S-TAB`   Re-align, move to previous field.

`RET`   Re-align the table and move down to next row. Creates a new row if necessary.

Column and row editing

`M-left`
`M-right`   Move the current column left/right.

`M-S-left`   Kill the current column.

`M-S-right`   Insert a new column to the left of the cursor position.

Chapter 3: Tables

M-up	
M-down	Move the current row up/down.
M-S-up	Kill the current row or horizontal line.
M-S-down	Insert a new row above the current row. With a prefix argument, the line is created below the current one.
C-c -	Insert a horizontal line below current row. With a prefix argument, the line is created above the current line.
C-c RET	Insert a horizontal line below current row, and move the cursor into the row below that line.
C-c ^	Sort the table lines in the region. The position of point indicates the column to be used for sorting, and the range of lines is the range between the nearest horizontal separator lines, or the entire table.

Further reading
Chapter 3 of the manual
Bastien's table tutorial
Bastien's spreadsheet tutorial
Eric's plotting tutorial

4 Hyperlinks

Like HTML, Org provides links inside a file, external links to other files, Usenet articles, emails, and much more.

4.1 Link format

Org will recognize plain URL-like links and activate them as clickable links. The general link format, however, looks like this:

    [[link][description]]   or alternatively   [[link]]

Once a link in the buffer is complete (all brackets present), Org will change the display so that 'description' is displayed instead of '[[link][description]]' and 'link' is displayed instead of '[[link]]'. To edit the invisible 'link' part, use `C-c C-l` with the cursor on the link.

4.2 Internal links

If the link does not look like a URL, it is considered to be internal in the current file. The most important case is a link like '[[#my-custom-id]]' which will link to the entry with the `CUSTOM_ID` property 'my-custom-id'.

Links such as '[[My Target]]' or '[[My Target][Find my target]]' lead to a text search in the current file for the corresponding target which looks like '<<My Target>>'.

Internal links will be used to reference their destination, through links or numbers, when possible.

4.3 External links

Org supports links to files, websites, Usenet and email messages, BBDB database entries and links to both IRC conversations and their logs. External links are URL-like locators. They start with a short identifying string followed by a colon. There can be no space after the colon. Here are some examples:

    http://www.astro.uva.nl/~dominik   on the web
    file:/home/dominik/images/jupiter.jpg   file, absolute path
    /home/dominik/images/jupiter.jpg   same as above
    file:papers/last.pdf   file, relative path
    file:projects.org   another Org file
    docview:papers/last.pdf::NNN   open file in doc-view mode at page NNN
    id:B7423F4D-2E8A-471B-8810-C40F074717E9   Link to heading by ID
    news:comp.emacs   Usenet link
    mailto:adent@galaxy.net   Mail link
    vm:folder   VM folder link
    vm:folder#id   VM message link
    wl:folder#id   WANDERLUST message link
    mhe:folder#id   MH-E message link
    rmail:folder#id   RMAIL message link
    gnus:group#id   Gnus article link
    bbdb:R.*Stallman   BBDB link (with regexp)
    irc:/irc.com/#emacs/bob   IRC link
    info:org:External%20links   Info node link (with encoded space)

A link should be enclosed in double brackets and may contain a descriptive text to be displayed instead of the URL (see Section 4.1 [Link format], page 8), for example:

Chapter 4: Hyperlinks

[[http://www.gnu.org/software/emacs/][GNU Emacs]]

If the description is a file name or URL that points to an image, HTML export (see Section 12.4 [HTML export], page 29) will inline the image as a clickable button. If there is no description at all and the link points to an image, that image will be inlined into the exported HTML file.

4.4 Handling links

Org provides methods to create a link in the correct syntax, to insert it into an Org file, and to follow the link.

C-c l   Store a link to the current location. This is a *global* command (you must create the key binding yourself) which can be used in any buffer to create a link. The link will be stored for later insertion into an Org buffer (see below).

C-c C-l   Insert a link. This prompts for a link to be inserted into the buffer. You can just type a link, or use history keys up and down to access stored links. You will be prompted for the description part of the link. When called with a C-u prefix argument, file name completion is used to link to a file.

C-c C-l (with cursor on existing link)
When the cursor is on an existing link, C-c C-l allows you to edit the link and description parts of the link.

C-c C-o or mouse-1 or mouse-2
Open link at point.

C-c &   Jump back to a recorded position. A position is recorded by the commands following internal links, and by C-c %. Using this command several times in direct succession moves through a ring of previously recorded positions.

4.5 Targeted links

File links can contain additional information to make Emacs jump to a particular location in the file when following a link. This can be a line number or a search option after a double colon.

Here is the syntax of the different ways to attach a search to a file link, together with an explanation:

[[file:~/code/main.c::255]]   Find line 255
[[file:~/xx.org::My Target]]   Find '<<My Target>>'
[[file:~/xx.org::#my-custom-id]]   Find entry with custom id

Further reading
Chapter 4 of the manual

5 TODO Items

Org mode does not require TODO lists to live in separate documents. Instead, TODO items are part of a notes file, because TODO items usually come up while taking notes! With Org mode, simply mark any entry in a tree as being a TODO item. In this way, information is not duplicated, and TODO items remain in the context from which they emerged.

Org mode provides methods to give you an overview of all the things that you have to do, collected from many files.

5.1 Using TODO states

Any headline becomes a TODO item when it starts with the word 'TODO', for example:
```
*** TODO Write letter to Sam Fortune
```
The most important commands to work with TODO entries are:

C-c C-t   Rotate the TODO state of the current item among
           (unmarked) -> TODO -> DONE -> (unmarked)

           The same rotation can also be done "remotely" from the agenda buffers with the t command key (see Section 10.4 [Agenda commands], page 23).

S-right/left
           Select the following/preceding TODO state, similar to cycling.

C-c / t   View TODO items in a *sparse tree* (see Section 2.6 [Sparse trees], page 4). Folds the buffer, but shows all TODO items and the headings hierarchy above them.

C-c a t   Show the global TODO list. Collects the TODO items from all agenda files (see Chapter 10 [Agenda Views], page 21) into a single buffer. See Section 10.3.2 [Global TODO list], page 22, for more information.

S-M-RET   Insert a new TODO entry below the current one.

Changing a TODO state can also trigger tag changes. See the docstring of the option org-todo-state-tags-triggers for details.

5.2 Multi-state workflows

You can use TODO keywords to indicate *sequential* working progress states:
```
(setq org-todo-keywords
      '((sequence "TODO" "FEEDBACK" "VERIFY" "|" "DONE" "DELEGATED")))
```
The vertical bar separates the TODO keywords (states that *need action*) from the DONE states (which need *no further action*). If you don't provide the separator bar, the last state is used as the DONE state. With this setup, the command C-c C-t will cycle an entry from TODO to FEEDBACK, then to VERIFY, and finally to DONE and DELEGATED. Sometimes you may want to use different sets of TODO keywords in parallel. For example, you may want to have the basic TODO/DONE, but also a workflow for bug fixing. Your setup would then look like this:
```
(setq org-todo-keywords
      '((sequence "TODO(t)" "|" "DONE(d)")
        (sequence "REPORT(r)" "BUG(b)" "KNOWNCAUSE(k)" "|" "FIXED(f)")))
```

Chapter 5: TODO Items

The keywords should all be different, this helps Org mode to keep track of which subsequence should be used for a given entry. The example also shows how to define keys for fast access of a particular state, by adding a letter in parenthesis after each keyword—you will be prompted for the key after C-c C-t.

To define TODO keywords that are valid only in a single file, use the following text anywhere in the file.

```
#+TODO: TODO(t) | DONE(d)
#+TODO: REPORT(r) BUG(b) KNOWNCAUSE(k) | FIXED(f)
#+TODO: | CANCELED(c)
```

After changing one of these lines, use C-c C-c with the cursor still in the line to make the changes known to Org mode.

5.3 Progress logging

Org mode can automatically record a timestamp and possibly a note when you mark a TODO item as DONE, or even each time you change the state of a TODO item. This system is highly configurable; settings can be on a per-keyword basis and can be localized to a file or even a subtree. For information on how to clock working time for a task, see Section 8.4 [Clocking work time], page 18.

Closing items

The most basic logging is to keep track of *when* a certain TODO item was finished. This is achieved with[1].

```
(setq org-log-done 'time)
```

Then each time you turn an entry from a TODO (not-done) state into any of the DONE states, a line 'CLOSED: [timestamp]' will be inserted just after the headline. If you want to record a note along with the timestamp, use[2]

```
(setq org-log-done 'note)
```

You will then be prompted for a note, and that note will be stored below the entry with a 'Closing Note' heading.

Tracking TODO state changes

You might want to keep track of TODO state changes. You can either record just a timestamp, or a time-stamped note for a change. These records will be inserted after the headline as an itemized list. When taking a lot of notes, you might want to get the notes out of the way into a drawer. Customize the variable org-log-into-drawer to get this behavior.

For state logging, Org mode expects configuration on a per-keyword basis. This is achieved by adding special markers '!' (for a timestamp) and '@' (for a note) in parentheses after each keyword. For example:

```
#+TODO: TODO(t) WAIT(w@/!) | DONE(d!) CANCELED(c@)
```

will define TODO keywords and fast access keys, and also request that a time is recorded when the entry is set to DONE, and that a note is recorded when switching to WAIT or CANCELED. The same syntax works also when setting org-todo-keywords.

---

[1] The corresponding in-buffer setting is: #+STARTUP: logdone
[2] The corresponding in-buffer setting is: #+STARTUP: lognotedone

Chapter 5: TODO Items

5.4 Priorities

If you use Org mode extensively, you may end up with enough TODO items that it starts to make sense to prioritize them. Prioritizing can be done by placing a *priority cookie* into the headline of a TODO item, like this

```
*** TODO [#A] Write letter to Sam Fortune
```

Org mode supports three priorities: 'A', 'B', and 'C'. 'A' is the highest, 'B' the default if none is given. Priorities make a difference only in the agenda.

C-c ,   Set the priority of the current headline. Press 'A', 'B' or 'C' to select a priority, or SPC to remove the cookie.

S-up/dwn   Increase/decrease priority of current headline

5.5 Breaking tasks down into subtasks

It is often advisable to break down large tasks into smaller, manageable subtasks. You can do this by creating an outline tree below a TODO item, with detailed subtasks on the tree. To keep the overview over the fraction of subtasks that are already completed, insert either '[/]' or '[%]' anywhere in the headline. These cookies will be updated each time the TODO status of a child changes, or when pressing C-c C-c on the cookie. For example:

```
* Organize Party [33%]
** TODO Call people [1/2]
*** TODO Peter
*** DONE Sarah
** TODO Buy food
** DONE Talk to neighbor
```

5.6 Checkboxes

Every item in a plain list (see Section 2.7 [Plain lists], page 4) can be made into a checkbox by starting it with the string '[ ]'. Checkboxes are not included in the global TODO list, so they are often great to split a task into a number of simple steps. Here is an example of a checkbox list.

```
* TODO Organize party [1/3]
  - [-] call people [1/2]
    - [ ] Peter
    - [X] Sarah
  - [X] order food
```

Checkboxes work hierarchically, so if a checkbox item has children that are checkboxes, toggling one of the children checkboxes will make the parent checkbox reflect if none, some, or all of the children are checked.

The following commands work with checkboxes:

C-c C-c   Toggle checkbox status or (with prefix arg) checkbox presence at point.

M-S-RET   Insert a new item with a checkbox. This works only if the cursor is already in a plain list item (see Section 2.7 [Plain lists], page 4).

Further reading
Chapter 5 of the manual
David O'Toole's introductory tutorial
Charles Cave's GTD setup

6 Tags

An excellent way to implement labels and contexts for cross-correlating information is to assign *tags* to headlines. Org mode has extensive support for tags.

Every headline can contain a list of tags; they occur at the end of the headline. Tags are normal words containing letters, numbers, '_', and '@'. Tags must be preceded and followed by a single colon, e.g., ':work:'. Several tags can be specified, as in ':work:urgent:'. Tags will by default be in bold face with the same color as the headline.

6.1 Tag inheritance

Tags make use of the hierarchical structure of outline trees. If a heading has a certain tag, all subheadings will inherit the tag as well. For example, in the list

```
* Meeting with the French group      :work:
** Summary by Frank                  :boss:notes:
*** TODO Prepare slides for him      :action:
```

the final heading will have the tags ':work:', ':boss:', ':notes:', and ':action:' even though the final heading is not explicitly marked with those tags. You can also set tags that all entries in a file should inherit just as if these tags were defined in a hypothetical level zero that surrounds the entire file. Use a line like this[1]:

```
#+FILETAGS: :Peter:Boss:Secret:
```

6.2 Setting tags

Tags can simply be typed into the buffer at the end of a headline. After a colon, `M-TAB` offers completion on tags. There is also a special command for inserting tags:

`C-c C-q`   Enter new tags for the current headline. Org mode will either offer completion or a special single-key interface for setting tags, see below. After pressing `RET`, the tags will be inserted and aligned to `org-tags-column`. When called with a `C-u` prefix, all tags in the current buffer will be aligned to that column, just to make things look nice.

`C-c C-c`   When the cursor is in a headline, this does the same as `C-c C-q`.

Org will support tag insertion based on a *list of tags*. By default this list is constructed dynamically, containing all tags currently used in the buffer. You may also globally specify a hard list of tags with the variable `org-tag-alist`. Finally you can set the default tags for a given file with lines like

```
#+TAGS: @work @home @tennisclub
#+TAGS: laptop car pc sailboat
```

By default Org mode uses the standard minibuffer completion facilities for entering tags. However, it also implements another, quicker, tag selection method called *fast tag selection*. This allows you to select and deselect tags with just a single key press. For this to work well you should assign unique letters to most of your commonly used tags. You can do this globally by configuring the variable `org-tag-alist` in your `.emacs` file. For example, you may find the need to tag many items in different files with ':@home:'. In this case you can set something like:

---

[1] As with all these in-buffer settings, pressing `C-c C-c` activates any changes in the line.

```
(setq org-tag-alist '(("@work" . ?w) ("@home" . ?h) ("laptop" . ?l)))
```
If the tag is only relevant to the file you are working on, then you can instead set the TAGS option line as:
```
#+TAGS: @work(w) @home(h) @tennisclub(t) laptop(l) pc(p)
```

6.3 Tag groups

In a set of mutually exclusive tags, the first tag can be defined as a *group tag*. When you search for a group tag, it will return matches for all members in the group. In an agenda view, filtering by a group tag will display headlines tagged with at least one of the members of the group. This makes tag searches and filters even more flexible.

You can set group tags by inserting a colon between the group tag and other tags, like this:
```
#+TAGS: { @read : @read_book  @read_ebook }
```
In this example, '@read' is a *group tag* for a set of three tags: '@read', '@read_book' and '@read_ebook'.

You can also use the :grouptags keyword directly when setting *org-tag-alist*, see the documentation of that variable.

If you want to ignore group tags temporarily, toggle group tags support with `org-toggle-tags-groups`, bound to `C-c C-x q`. If you want to disable tag groups completely, set *org-group-tags* to nil.

6.4 Tag searches

Once a system of tags has been set up, it can be used to collect related information into special lists.

`C-c \`
`C-c / m`   Create a sparse tree with all headlines matching a tags search. With a `C-u` prefix argument, ignore headlines that are not a TODO line.

`C-c a m`   Create a global list of tag matches from all agenda files. See Section 10.3.3 [Matching tags and properties], page 22.

`C-c a M`   Create a global list of tag matches from all agenda files, but check only TODO items and force checking subitems (see variable `org-tags-match-list-sublevels`).

These commands all prompt for a match string which allows basic Boolean logic like '+boss+urgent-project1', to find entries with tags 'boss' and 'urgent', but not 'project1', or 'Kathy|Sally' to find entries which are tagged, like 'Kathy' or 'Sally'. The full syntax of the search string is rich and allows also matching against TODO keywords, entry levels and properties. For a complete description with many examples, see Section 10.3.3 [Matching tags and properties], page 22.

Further reading
Chapter 6 of the manual
Sacha Chua's article about tagging in Org-mode

7 Properties

Properties are key-value pairs associated with an entry. They live in a special drawer with the name `PROPERTIES`. Each property is specified on a single line, with the key (surrounded by colons) first, and the value after it:

```
* CD collection
** Classic
*** Goldberg Variations
    :PROPERTIES:
    :Title:     Goldberg Variations
    :Composer:  J.S. Bach
    :Publisher: Deutsche Grammophon
    :NDisks:    1
    :END:
```

You may define the allowed values for a particular property ':Xyz:' by setting a property ':Xyz_ALL:'. This special property is *inherited*, so if you set it in a level 1 entry, it will apply to the entire tree. When allowed values are defined, setting the corresponding property becomes easier and is less prone to typing errors. For the example with the CD collection, we can predefine publishers and the number of disks in a box like this:

```
* CD collection
  :PROPERTIES:
  :NDisks_ALL:    1 2 3 4
  :Publisher_ALL: "Deutsche Grammophon" Philips EMI
  :END:
```

or globally using `org-global-properties`, or file-wide like this:

```
#+PROPERTY: NDisks_ALL 1 2 3 4
```

C-c C-x p Set a property. This prompts for a property name and a value.

C-c C-c d Remove a property from the current entry.

To create sparse trees and special lists with selection based on properties, the same commands are used as for tag searches (see Section 6.4 [Tag searches], page 14). The syntax for the search string is described in Section 10.3.3 [Matching tags and properties], page 22.

Further reading
Chapter 7 of the manual
Bastien's column view tutorial

8 Dates and Times

To assist project planning, TODO items can be labeled with a date and/or a time. The specially formatted string carrying the date and time information is called a *timestamp* in Org mode.

8.1 Timestamps

A timestamp is a specification of a date (possibly with a time or a range of times) in a special format, either '<2003-09-16 Tue>' or '<2003-09-16 Tue 09:39>' or '<2003-09-16 Tue 12:00-12:30>'. A timestamp can appear anywhere in the headline or body of an Org tree entry. Its presence causes entries to be shown on specific dates in the agenda (see Section 10.3.1 [Weekly/daily agenda], page 21). We distinguish:

Plain timestamp; Event; Appointment
A simple timestamp just assigns a date/time to an item. This is just like writing down an appointment or event in a paper agenda.

```
* Meet Peter at the movies
  <2006-11-01 Wed 19:15>
* Discussion on climate change
  <2006-11-02 Thu 20:00-22:00>
```

Timestamp with repeater interval
A timestamp may contain a *repeater interval*, indicating that it applies not only on the given date, but again and again after a certain interval of N days (d), weeks (w), months (m), or years (y). The following will show up in the agenda every Wednesday:

```
* Pick up Sam at school
  <2007-05-16 Wed 12:30 +1w>
```

Diary-style sexp entries
For more complex date specifications, Org mode supports using the special sexp diary entries implemented in the Emacs calendar/diary package. For example

```
* The nerd meeting on every 2nd Thursday of the month
  <%%(diary-float t 4 2)>
```

Time/Date range
Two timestamps connected by '--' denote a range.

```
** Meeting in Amsterdam
   <2004-08-23 Mon>--<2004-08-26 Thu>
```

Inactive timestamp
Just like a plain timestamp, but with square brackets instead of angular ones. These timestamps are inactive in the sense that they do *not* trigger an entry to show up in the agenda.

```
* Gillian comes late for the fifth time
  [2006-11-01 Wed]
```

8.2 Creating timestamps

For Org mode to recognize timestamps, they need to be in the specific format. All commands listed below produce timestamps in the correct format.

Chapter 8: Dates and Times

`C-c .` Prompt for a date and insert a corresponding timestamp. When the cursor is at an existing timestamp in the buffer, the command is used to modify this timestamp instead of inserting a new one. When this command is used twice in succession, a time range is inserted. With a prefix, also add the current time.

`C-c !` Like `C-c .`, but insert an inactive timestamp that will not cause an agenda entry.

`S-left/right`
Change date at cursor by one day.

`S-up/down`
Change the item under the cursor in a timestamp. The cursor can be on a year, month, day, hour or minute. When the timestamp contains a time range like '15:30-16:30', modifying the first time will also shift the second, shifting the time block with constant length. To change the length, modify the second time.

When Org mode prompts for a date/time, it will accept any string containing some date and/or time information, and intelligently interpret the string, deriving defaults for unspecified information from the current date and time. You can also select a date in the pop-up calendar. See the manual for more information on how exactly the date/time prompt works.

8.3 Deadlines and scheduling

A timestamp may be preceded by special keywords to facilitate planning:

DEADLINE
Meaning: the task (most likely a TODO item, though not necessarily) is supposed to be finished on that date.

`C-c C-d` Insert 'DEADLINE' keyword along with a stamp, in the line following the headline.

On the deadline date, the task will be listed in the agenda. In addition, the agenda for *today* will carry a warning about the approaching or missed deadline, starting `org-deadline-warning-days` before the due date, and continuing until the entry is marked DONE. An example:

```
*** TODO write article about the Earth for the Guide
    The editor in charge is [[bbdb:Ford Prefect]]
    DEADLINE: <2004-02-29 Sun>
```

SCHEDULED
Meaning: you are *planning to start working* on that task on the given date[1].

`C-c C-s` Insert 'SCHEDULED' keyword along with a stamp, in the line following the headline.

The headline will be listed under the given date[2]. In addition, a reminder that the scheduled date has passed will be present in the compilation for *today*, until the entry is marked DONE. I.e. the task will automatically be forwarded until completed.

---

[1] This is quite different from what is normally understood by *scheduling a meeting*, which is done in Org-mode by just inserting a time stamp without keyword.

[2] It will still be listed on that date after it has been marked DONE. If you don't like this, set the variable `org-agenda-skip-scheduled-if-done`.

Chapter 8: Dates and Times

```
*** TODO Call Trillian for a date on New Years Eve.
    SCHEDULED: <2004-12-25 Sat>
```

Some tasks need to be repeated again and again. Org mode helps to organize such tasks using a so-called repeater in a DEADLINE, SCHEDULED, or plain timestamp. In the following example

```
** TODO Pay the rent
   DEADLINE: <2005-10-01 Sat +1m>
```

the +1m is a repeater; the intended interpretation is that the task has a deadline on <2005-10-01> and repeats itself every (one) month starting from that time.

8.4 Clocking work time

Org mode allows you to clock the time you spend on specific tasks in a project.

C-c C-x C-i
: Start the clock on the current item (clock-in). This inserts the CLOCK keyword together with a timestamp. When called with a C-u prefix argument, select the task from a list of recently clocked tasks.

C-c C-x C-o
: Stop the clock (clock-out). This inserts another timestamp at the same location where the clock was last started. It also directly computes the resulting time in inserts it after the time range as '=> HH:MM'.

C-c C-x C-e
: Update the effort estimate for the current clock task.

C-c C-x C-q
: Cancel the current clock. This is useful if a clock was started by mistake, or if you ended up working on something else.

C-c C-x C-j
: Jump to the entry that contains the currently running clock. With a C-u prefix arg, select the target task from a list of recently clocked tasks.

C-c C-x C-r
: Insert a dynamic block containing a clock report as an Org-mode table into the current file. When the cursor is at an existing clock table, just update it.
  ```
  #+BEGIN: clocktable :maxlevel 2 :emphasize nil :scope file
  #+END: clocktable
  ```
  For details about how to customize this view, see the manual.

C-c C-c
: Update dynamic block at point. The cursor needs to be in the #+BEGIN line of the dynamic block.

The l key may be used in the agenda (see Section 10.3.1 [Weekly/daily agenda], page 21) to show which tasks have been worked on or closed during a day.

Further reading
Chapter 8 of the manual
Charles Cave's Date and Time tutorial
Bernt Hansen's clocking workflow

9 Capture - Refile - Archive

An important part of any organization system is the ability to quickly capture new ideas and tasks, and to associate reference material with them. Org defines a capture process to create tasks. Once in the system, tasks and projects need to be moved around. Moving completed project trees to an archive file keeps the system compact and fast.

9.1 Capture

Org's lets you store quick notes with little interruption of your work flow. You can define templates for new entries and associate them with different targets for storing notes.

Setting up a capture location

The following customization sets a default target[1] file for notes, and defines a global key for capturing new stuff.

```
(setq org-default-notes-file (concat org-directory "/notes.org"))
(define-key global-map "\C-cc" 'org-capture)
```

Using capture

`C-c c` Start a capture process, placing you into a narrowed indirect buffer to edit.

`C-c C-c` Once you are done entering information into the capture buffer, `C-c C-c` will return you to the window configuration before the capture process, so that you can resume your work without further distraction.

`C-c C-w` Finalize by moving the entry to a refile location (see section 9.2).

`C-c C-k` Abort the capture process and return to the previous state.

Capture templates

You can use templates to generate different types of capture notes, and to store them in different places. For example, if you would like to store new tasks under a heading 'Tasks' in file `TODO.org`, and journal entries in a date tree in `journal.org` you could use:

```
(setq org-capture-templates
 '(("t" "Todo" entry (file+headline "~/org/gtd.org" "Tasks")
        "* TODO %?\n  %i\n  %a")
   ("j" "Journal" entry (file+datetree "~/org/journal.org")
        "* %?\nEntered on %U\n  %i\n  %a")))
```

In these entries, the first string is the key to reach the template, the second is a short description. Then follows the type of the entry and a definition of the target location for storing the note. Finally, the template itself, a string with %-escapes to fill in information based on time and context.

When you call `M-x org-capture`, Org will prompt for a key to select the template (if you have more than one template) and then prepare the buffer like

```
* TODO
  [[file:link to where you were when initiating capture]]
```

---

[1] Using capture templates, you get finer control over capture locations, see [Capture templates], page 19.

During expansion of the template, special %-escapes[2] allow dynamic insertion of content. Here is a small selection of the possibilities, consult the manual for more.

%a	annotation, normally the link created with `org-store-link`
%i	initial content, the region when capture is called with C-u.
%t, %T	timestamp, date only, or date and time
%u, %U	like above, but inactive timestamps

9.2 Refile and copy

When reviewing the captured data, you may want to refile or copy some of the entries into a different list, for example into a project. Cutting, finding the right location, and then pasting the note is cumbersome. To simplify this process, use the following commands:

`C-c M-w` Copy the entry or region at point. This command behaves like `org-refile`, except that the original note will not be deleted.

`C-c C-w` Refile the entry or region at point. This command offers possible locations for refiling the entry and lets you select one with completion. The item (or all items in the region) is filed below the target heading as a subitem.
By default, all level 1 headlines in the current buffer are considered to be targets, but you can have more complex definitions across a number of files. See the variable `org-refile-targets` for details.

`C-u C-c C-w`
Use the refile interface to jump to a heading.

`C-u C-u C-c C-w`
Jump to the location where `org-refile` last moved a tree to.

9.3 Archiving

When a project represented by a (sub)tree is finished, you may want to move the tree out of the way and to stop it from contributing to the agenda. Archiving is important to keep your working files compact and global searches like the construction of agenda views fast. The most common archiving action is to move a project tree to another file, the archive file.

`C-c C-x C-a`
Archive the current entry using `org-archive-default-command`.

`C-c C-x C-s` or short `C-c $`
Archive the subtree starting at the cursor position to the location given by `org-archive-location`.

The default archive location is a file in the same directory as the current file, with the name derived by appending _archive to the current file name. For information and examples on how to change this, see the documentation string of the variable `org-archive-location`. There is also an in-buffer option for setting this variable, for example

    #+ARCHIVE: %s_done::

Further reading
Chapter 9 of the manual
Sebastian Rose's tutorial for capturing from a web browser

---

[2] If you need one of these sequences literally, escape the % with a backslash.

10 Agenda Views

Due to the way Org works, TODO items, time-stamped items, and tagged headlines can be scattered throughout a file or even a number of files. To get an overview of open action items, or of events that are important for a particular date, this information must be collected, sorted and displayed in an organized way. There are several different views, see below.

The extracted information is displayed in a special *agenda buffer*. This buffer is read-only, but provides commands to visit the corresponding locations in the original Org files, and even to edit these files remotely. Remote editing from the agenda buffer means, for example, that you can change the dates of deadlines and appointments from the agenda buffer. The commands available in the Agenda buffer are listed in Section 10.4 [Agenda commands], page 23.

10.1 Agenda files

The information to be shown is normally collected from all *agenda files*, the files listed in the variable `org-agenda-files`.

C-c [Add current file to the list of agenda files. The file is added to the front of the list. If it was already in the list, it is moved to the front. With a prefix argument, file is added/moved to the end.

C-c]   Remove current file from the list of agenda files.

C-,   Cycle through agenda file list, visiting one file after the other.

10.2 The agenda dispatcher

The views are created through a dispatcher, which should be bound to a global key—for example C-c a (see Section 1.2 [Installation], page 1). After pressing C-c a, an additional letter is required to execute a command:

a   The calendar-like agenda (see Section 10.3.1 [Weekly/daily agenda], page 21).

t / T   A list of all TODO items (see Section 10.3.2 [Global TODO list], page 22).

m / M   A list of headlines matching a TAGS expression (see Section 10.3.3 [Matching tags and properties], page 22).

s   A list of entries selected by a boolean expression of keywords and/or regular expressions that must or must not occur in the entry.

10.3 The built-in agenda views

10.3.1 The weekly/daily agenda

The purpose of the weekly/daily *agenda* is to act like a page of a paper agenda, showing all the tasks for the current week or day.

C-c a a   Compile an agenda for the current week from a list of Org files. The agenda shows the entries for each day.

Emacs contains the calendar and diary by Edward M. Reingold. Org-mode understands the syntax of the diary and allows you to use diary sexp entries directly in Org files:

Chapter 10: Agenda Views

```
* Birthdays and similar stuff
#+CATEGORY: Holiday
%%(org-calendar-holiday)   ; special function for holiday names
#+CATEGORY: Ann
%%(diary-anniversary  5 14 1956)[1] Arthur Dent is %d years old
%%(diary-anniversary 10  2 1869) Mahatma Gandhi would be %d years old
```

Org can interact with Emacs appointments notification facility. To add all the appointments of your agenda files, use the command `org-agenda-to-appt`. See the docstring for details.

10.3.2 The global TODO list

The global TODO list contains all unfinished TODO items formatted and collected into a single place. Remote editing of TODO items lets you can change the state of a TODO entry with a single key press. The commands available in the TODO list are described in Section 10.4 [Agenda commands], page 23.

`C-c a t` Show the global TODO list. This collects the TODO items from all agenda files (see Chapter 10 [Agenda Views], page 21) into a single buffer.

`C-c a T` Like the above, but allows selection of a specific TODO keyword.

10.3.3 Matching tags and properties

If headlines in the agenda files are marked with *tags* (see Chapter 6 [Tags], page 13), or have properties (see Chapter 7 [Properties], page 15), you can select headlines based on this metadata and collect them into an agenda buffer. The match syntax described here also applies when creating sparse trees with `C-c / m`. The commands available in the tags list are described in Section 10.4 [Agenda commands], page 23.

`C-c a m` Produce a list of all headlines that match a given set of tags. The command prompts for a selection criterion, which is a boolean logic expression with tags, like '+work+urgent-withboss' or 'work|home' (see Chapter 6 [Tags], page 13). If you often need a specific search, define a custom command for it (see Section 10.2 [Agenda dispatcher], page 21).

`C-c a M` Like `C-c a m`, but only select headlines that are also TODO items.

Match syntax

A search string can use Boolean operators '&' for AND and '|' for OR. '&' binds more strongly than '|'. Parentheses are currently not implemented. Each element in the search is either a tag, a regular expression matching tags, or an expression like PROPERTY OPERATOR VALUE with a comparison operator, accessing a property value. Each element may be preceded by '-', to select against it, and '+' is syntactic sugar for positive selection. The AND operator '&' is optional when '+' or '-' is present. Here are some examples, using only tags.

'+work-boss'
     Select headlines tagged ':work:', but discard those also tagged ':boss:'.

'work|laptop'
     Selects lines tagged ':work:' or ':laptop:'.

---

[1] Note that the order of the arguments (month, day, year) depends on the setting of `calendar-date-style`.

'work|laptop+night'
: Like before, but require the ':laptop:' lines to be tagged also ':night:'.

You may also test for properties at the same time as matching tags, see the manual for more information.

10.3.4 Search view

This agenda view is a general text search facility for Org mode entries. It is particularly useful to find notes.

C-c a s
: This is a special search that lets you select entries by matching a substring or specific words using a boolean logic.

For example, the search string 'computer equipment' will find entries that contain 'computer equipment' as a substring. Search view can also search for specific keywords in the entry, using Boolean logic. The search string '+computer +wifi -ethernet -{8\.11[bg]}' will search for note entries that contain the keywords computer and wifi, but not the keyword ethernet, and which are also not matched by the regular expression 8\.11[bg], meaning to exclude both 8.11b and 8.11g.

Note that in addition to the agenda files, this command will also search the files listed in org-agenda-text-search-extra-files.

10.4 Commands in the agenda buffer

Entries in the agenda buffer are linked back to the Org file or diary file where they originate. Commands are provided to show and jump to the original entry location, and to edit the Org files "remotely" from the agenda buffer. This is just a selection of the many commands, explore the Agenda menu and the manual for a complete list.

Motion

n
: Next line (same as up and C-p).

p
: Previous line (same as down and C-n).

View/Go to Org file

mouse-3

SPC
: Display the original location of the item in another window. With prefix arg, make sure that the entire entry is made visible in the outline, not only the heading.

TAB
: Go to the original location of the item in another window. Under Emacs 22, mouse-1 will also work for this.

RET
: Go to the original location of the item and delete other windows.

Change display

o
: Delete other windows.

d / w
: Switch to day/week view.

f and b
: Go forward/backward in time to display the following org-agenda-current-span days. For example, if the display covers a week, switch to the following/previous week.

Chapter 10: Agenda Views

.	Go to today.
j	Prompt for a date and go there.
v l or short l	
	Toggle Logbook mode. In Logbook mode, entries that were marked DONE while logging was on (variable `org-log-done`) are shown in the agenda, as are entries that have been clocked on that day. When called with a `C-u` prefix, show all possible logbook entries, including state changes.
r or g	Recreate the agenda buffer, to reflect the changes.
s	Save all Org buffers in the current Emacs session, and also the locations of IDs.

Secondary filtering and query editing

/	Filter the current agenda view with respect to a tag. You are prompted for a letter to select a tag. Press '-' first to select against the tag.
\	Narrow the current agenda filter by an additional condition.

Remote editing (see the manual for many more commands)

0--9	Digit argument.
t	Change the TODO state of the item, in the agenda and in the org file.
C-k	Delete the current agenda item along with the entire subtree belonging to it in the original Org file.
C-c C-w	Refile the entry at point.
C-c C-x C-a or short a	
	Archive the subtree corresponding to the entry at point using the default archiving command set in `org-archive-default-command`.
C-c C-x C-s or short $	
	Archive the subtree corresponding to the current headline.
C-c C-s	Schedule this item, with prefix arg remove the scheduling timestamp
C-c C-d	Set a deadline for this item, with prefix arg remove the deadline.
S-right and S-left	
	Change the timestamp associated with the current line by one day.
I	Start the clock on the current item.
O / X	Stop/cancel the previously started clock.
J	Jump to the running clock in another window.

10.5 Custom agenda views

The main application of custom searches is the definition of keyboard shortcuts for frequently used searches, either creating an agenda buffer, or a sparse tree (the latter covering of course only the current buffer). Custom commands are configured in the variable `org-agenda-custom-commands`. You can customize this variable, for example by pressing `C-c a C`. You can also directly set it with Emacs Lisp in `.emacs`. The following example contains all valid search types:

Chapter 10: Agenda Views

```
(setq org-agenda-custom-commands
      '(("w" todo "WAITING")
        ("u" tags "+boss-urgent")
        ("v" tags-todo "+boss-urgent")))
```

The initial string in each entry defines the keys you have to press after the dispatcher command `C-c a` in order to access the command. Usually this will be just a single character. The second parameter is the search type, followed by the string or regular expression to be used for the matching. The example above will therefore define:

`C-c a w`   as a global search for TODO entries with 'WAITING' as the TODO keyword

`C-c a u`   as a global tags search for headlines marked ':boss:' but not ':urgent:'

`C-c a v`   as the same search as `C-c a u`, but limiting the search to headlines that are also TODO items

Further reading
Chapter 10 of the manual
Mat Lundin's tutorial about custom agenda commands
John Wiegley's setup

11 Markup for rich export

When exporting Org-mode documents, the exporter tries to reflect the structure of the document as accurately as possible in the backend. Since export targets like HTML, LaTeX, or DocBook allow much richer formatting, Org mode has rules on how to prepare text for rich export. This section summarizes the markup rules used in an Org-mode buffer.

11.1 Structural markup elements

Document title

The title of the exported document is taken from the special line
```
#+TITLE: This is the title of the document
```

Headings and sections

The outline structure of the document as described in Chapter 2 [Document Structure], page 2, forms the basis for defining sections of the exported document. However, since the outline structure is also used for (for example) lists of tasks, only the first three outline levels will be used as headings. Deeper levels will become itemized lists. You can change the location of this switch globally by setting the variable `org-export-headline-levels`, or on a per-file basis with a line
```
#+OPTIONS: H:4
```

Table of contents

The table of contents is normally inserted directly before the first headline of the file.
```
#+OPTIONS: toc:2        (only to two levels in TOC)
#+OPTIONS: toc:nil      (no TOC at all)
```

Paragraphs, line breaks, and quoting

Paragraphs are separated by at least one empty line. If you need to enforce a line break within a paragraph, use '\\' at the end of a line.

To keep the line breaks in a region, but otherwise use normal formatting, you can use this construct, which can also be used to format poetry.
```
#+BEGIN_VERSE
 Great clouds overhead
 Tiny black birds rise and fall
 Snow covers Emacs

     -- AlexSchroeder
#+END_VERSE
```

When quoting a passage from another document, it is customary to format this as a paragraph that is indented on both the left and the right margin. You can include quotations in Org-mode documents like this:
```
#+BEGIN_QUOTE
Everything should be made as simple as possible,
but not any simpler -- Albert Einstein
#+END_QUOTE
```

If you would like to center some text, do it like this:

```
#+BEGIN_CENTER
Everything should be made as simple as possible, \\
but not any simpler
#+END_CENTER
```

Emphasis and monospace

You can make words *bold*, /italic/, _underlined_, =verbatim= and ~code~, and, if you must, '+strike-through+'. Text in the code and verbatim string is not processed for Org-mode specific syntax, it is exported verbatim. To insert a horizontal rules, use a line consisting of only dashes, and at least 5 of them.

Comment lines

Lines starting with zero or more whitespace characters followed by '#' and a whitespace are treated as comments and, as such, are not exported.

Likewise, regions surrounded by '#+BEGIN_COMMENT' ... '#+END_COMMENT' are not exported.

Finally, a 'COMMENT' keyword at the beginning of an entry, but after any other keyword or priority cookie, comments out the entire subtree. The command below helps changing the comment status of a headline.

C-c ;   Toggle the COMMENT keyword at the beginning of an entry.

11.2 Images and Tables

For Org mode tables, the lines before the first horizontal separator line will become table header lines. You can use the following lines somewhere before the table to assign a caption and a label for cross references, and in the text you can refer to the object with [[tab:basic-data]]:

```
#+CAPTION: This is the caption for the next table (or link)
#+NAME:    tbl:basic-data
   | ... | ...|
   |-----|----|
```

Some backends allow you to directly include images into the exported document. Org does this, if a link to an image files does not have a description part, for example [[./img/a.jpg]]. If you wish to define a caption for the image and maybe a label for internal cross references, you sure that the link is on a line by itself precede it with:

```
#+CAPTION: This is the caption for the next figure link (or table)
#+NAME:    fig:SED-HR4049
[[./img/a.jpg]]
```

The same caption mechanism applies to other structures than images and tables (e.g., LaTeX equations, source code blocks), provided the chosen export back-end supports them.

11.3 Literal examples

You can include literal examples that should not be subjected to markup. Such examples will be typeset in monospace, so this is well suited for source code and similar examples.

```
#+BEGIN_EXAMPLE
Some example from a text file.
#+END_EXAMPLE
```

Chapter 11: Markup for rich export

For simplicity when using small examples, you can also start the example lines with a colon followed by a space. There may also be additional whitespace before the colon:

```
Here is an example
  : Some example from a text file.
```

For source code from a programming language, or any other text that can be marked up by font-lock in Emacs, you can ask for it to look like the fontified Emacs buffer

```
#+BEGIN_SRC emacs-lisp
(defun org-xor (a b)
  "Exclusive or."
  (if a (not b) b))
#+END_SRC
```

To edit the example in a special buffer supporting this language, use `C-c '` to both enter and leave the editing buffer.

11.4 Include files

During export, you can include the content of another file. For example, to include your .emacs file, you could use:

```
#+INCLUDE: "~/.emacs" src emacs-lisp
```

The optional second and third parameter are the markup (i.e., 'example' or 'src'), and, if the markup is 'src', the language for formatting the contents. The markup is optional, if it is not given, the text will be assumed to be in Org mode format and will be processed normally. File-links will be interpreted as well:

```
#+INCLUDE: "./otherfile.org::#my_custom_id" :only-contents t
```

`C-c '` will visit the included file.

11.5 Embedded LaTeX

For scientific notes which need to be able to contain mathematical symbols and the occasional formula, Org-mode supports embedding LaTeX code into its files. You can directly use TeX-like syntax for special symbols, enter formulas and entire LaTeX environments.

```
Angles are written as Greek letters \alpha, \beta and \gamma. The mass if
the sun is M_sun = 1.989 x 10^30 kg. The radius of the sun is R_{sun} =
6.96 x 10^8 m. If $a^2=b$ and $b=2$, then the solution must be either
$a=+\sqrt{2}$ or $a=-\sqrt{2}$.

\begin{equation}
x=\sqrt{b}
\end{equation}
```

With special setup, LaTeX snippets will be included as images when exporting to HTML.

Further reading

Chapter 11 of the manual

12 Exporting

Org-mode documents can be exported into a variety of other formats: ASCII export for inclusion into emails, HTML to publish on the web, LaTeX/PDF for beautiful printed documents and DocBook to enter the world of many other formats using DocBook tools. There is also export to iCalendar format so that planning information can be incorporated into desktop calendars.

12.1 Export options

The exporter recognizes special lines in the buffer which provide additional information. These lines may be put anywhere in the file. The whole set of lines can be inserted into the buffer with `C-c C-e #`.

`C-c C-e #` Insert template with export options, see example below.

```
#+TITLE:     the title to be shown
#+AUTHOR:    the author (default taken from user-full-name)
#+DATE:      a date, fixed, or an Org timestamp
#+EMAIL:     his/her email address (default from user-mail-address)
#+LANGUAGE:  language, e.g. 'en' (org-export-default-language)
#+OPTIONS:   H:2 num:t toc:t \n:nil ::t |:t ^:t f:t tex:t ...
```

12.2 The export dispatcher

All export commands can be reached using the export dispatcher, which is a prefix key that prompts for an additional key specifying the command. Normally the entire file is exported, but if a region is active, it will be exported instead.

`C-c C-e` Dispatcher for export and publishing commands.

12.3 ASCII/Latin-1/UTF-8 export

ASCII export produces a simple and very readable version of an Org-mode file, containing only plain ASCII. Latin-1 and UTF-8 export augment the file with special characters and symbols available in these encodings.

`C-c C-e t a` and `C-c C-e t A`
   Export as ASCII file or temporary buffer.

`C-c C-e t n` and `C-c C-e t N`
   Like the above commands, but use Latin-1 encoding.

`C-c C-e t u` and `C-c C-e t U`
   Like the above commands, but use UTF-8 encoding.

12.4 HTML export

`C-c C-e h h`
   Export as HTML file `myfile.html`.

`C-c C-e h o`
   Export as HTML file and immediately open it with a browser.

To insert HTML that should be copied verbatim to the exported file use either
```
#+HTML: Literal HTML code for export
```
or
```
#+BEGIN_EXPORT html
All lines between these markers are exported literally
#+END_HTML
```

12.5 LaTeX and PDF export

`C-c C-e l l`
> Export as LaTeX file `myfile.tex`.

`C-c C-e l p`
> Export as LaTeX and then process to PDF.

`C-c C-e l o`
> Export as LaTeX and then process to PDF, then open the resulting PDF file.

By default, the LaTeX output uses the class `article`. You can change this by adding an option like `#+LATEX_CLASS: myclass` in your file. The class must be listed in `org-latex-classes`.

Embedded LaTeX as described in Section 11.5 [Embedded LaTeX], page 28, will be correctly inserted into the LaTeX file. Similarly to the HTML exporter, you can use `#+LATEX:` and `#+BEGIN_EXPORT latex ... #+END_EXPORT` construct to add verbatim LaTeX code.

12.6 iCalendar export

`C-c C-e c f`
> Create iCalendar entries for the current file in a `.ics` file.

`C-c C-e c c`
> Create a single large iCalendar file from all files in `org-agenda-files` and write it to the file given by `org-icalendar-combined-agenda-file`.

Further reading
Chapter 12 of the manual
Sebastian Rose's image handling tutorial
Thomas Dye's LaTeX export tutorial Eric Fraga's BEAMER presentation tutorial

13 Publishing

Org includes a publishing management system that allows you to configure automatic HTML conversion of *projects* composed of interlinked org files. You can also configure Org to automatically upload your exported HTML pages and related attachments, such as images and source code files, to a web server. For detailed instructions about setup, see the manual.

Here is an example:

```
(setq org-publish-project-alist
      '(("org"
         :base-directory "~/org/"
         :publishing-directory "~/public_html"
         :section-numbers nil
         :table-of-contents nil
         :style "<link rel=\"stylesheet\"
                 href=\"../other/mystyle.css\"
                 type=\"text/css\"/>")))
```

C-c C-e P x
: Prompt for a specific project and publish all files that belong to it.

C-c C-e P p
: Publish the project containing the current file.

C-c C-e P f
: Publish only the current file.

C-c C-e P a
: Publish every project.

Org uses timestamps to track when a file has changed. The above functions normally only publish changed files. You can override this and force publishing of all files by giving a prefix argument to any of the commands above.

Further reading
Chapter 13 of the manual
Sebastian Rose's publishing tutorial
Ian Barton's Jekyll/blogging setup

14 Working with source code

Org-mode provides a number of features for working with source code, including editing of code blocks in their native major-mode, evaluation of code blocks, tangling of code blocks, and exporting code blocks and their results in several formats.

Structure of Code Blocks
The structure of code blocks is as follows:
```
#+NAME: <name>
#+BEGIN_SRC <language> <switches> <header arguments>
  <body>
#+END_SRC
```
Where `<name>` is a string used to name the code block, `<language>` specifies the language of the code block (e.g. `emacs-lisp`, `shell`, `R`, `python`, etc...), `<switches>` can be used to control export of the code block, `<header arguments>` can be used to control many aspects of code block behavior as demonstrated below, and `<body>` contains the actual source code.

Editing source code
Use `C-c '` to edit the current code block. This brings up a language major-mode edit buffer containing the body of the code block. Saving this buffer will write the new contents back to the Org buffer. Use `C-c '` again to exit the edit buffer.

Evaluating code blocks
Use `C-c C-c` to evaluate the current code block and insert its results in the Org-mode buffer. By default, evaluation is only turned on for `emacs-lisp` code blocks, however support exists for evaluating blocks in many languages. For a complete list of supported languages see the manual. The following shows a code block and its results.
```
#+BEGIN_SRC emacs-lisp
  (+ 1 2 3 4)
#+END_SRC

#+RESULTS:
: 10
```

Extracting source code
Use `C-c C-v t` to create pure source code files by extracting code from source blocks in the current buffer. This is referred to as "tangling"—a term adopted from the literate programming community. During "tangling" of code blocks their bodies are expanded using `org-babel-expand-src-block` which can expand both variable and "noweb" style references. In order to tangle a code block it must have a `:tangle` header argument, see the manual for details.

Library of Babel
Use `C-c C-v l` to load the code blocks from an Org-mode files into the "Library of Babel", these blocks can then be evaluated from any Org-mode buffer. A collection of generally

Chapter 14: Working with source code

useful code blocks is accessible through Org-modes community-driven documentation on Worg.

Header Arguments

Many aspects of the evaluation and export of code blocks are controlled through header arguments. These can be specified globally, at the file level, at the outline subtree level, and at the individual code block level. The following describes some of the header arguments.

- `:var` The `:var` header argument is used to pass arguments to code blocks. The values passed to arguments can be literal values, values from org-mode tables and literal example blocks, or the results of other named code blocks.

- `:results` The `:results` header argument controls the *collection*, *type*, and *handling* of code block results. Values of `output` or `value` (the default) specify how results are collected from a code block's evaluation. Values of `vector`, `scalar file raw html latex` and `code` specify the type of the results of the code block which dictates how they will be incorporated into the Org-mode buffer. Values of `silent`, `replace`, `prepend`, and `append` specify handling of code block results, specifically if and how the results should be inserted into the Org-mode buffer.

- `:session` A header argument of `:session` will cause the code block to be evaluated in a persistent interactive inferior process in Emacs. This allows for persisting state between code block evaluations, and for manual inspection of the results of evaluation.

- `:exports` Any combination of the *code* or the *results* of a block can be retained on export, this is specified by setting the `:results` header argument to `code results none` or `both`.

- `:tangle` A header argument of `:tangle yes` will cause a code block's contents to be tangled to a file named after the filename of the Org-mode buffer. An alternate file name can be specified with `:tangle filename`.

- `:cache` A header argument of `:cache yes` will cause associate a hash of the expanded code block with the results, ensuring that code blocks are only re-run when their inputs have changed.

- `:noweb` A header argument of `:noweb yes` will expand "noweb" style references on evaluation and tangling.

- `:file` Code blocks which output results to files (e.g. graphs, diagrams and figures) can accept a `:file filename` header argument in which case the results are saved to the named file, and a link to the file is inserted into the Org-mode buffer.

Further reading
Chapter 11 and section 5 of the manual
The Babel site on Worg

15 Miscellaneous

15.1 Completion

Org supports in-buffer completion with `M-TAB`. This type of completion does not make use of the minibuffer. You simply type a few letters into the buffer and use the key to complete text right there. For example, this command will complete TeX symbols after '\', TODO keywords at the beginning of a headline, and tags after ':' in a headline.

15.2 A cleaner outline view

Some people find it noisy and distracting that the Org headlines start with a potentially large number of stars, and that text below the headlines is not indented. While this is no problem when writing a *book-like* document where the outline headings are really section headings, in a more *list-oriented* outline, indented structure is a lot cleaner:

```
* Top level headline          |    * Top level headline
** Second level               |      * Second level
*** 3rd level                 |        * 3rd level
some text                     |          some text
*** 3rd level                 |        * 3rd level
more text                     |          more text
* Another top level headline  |    * Another top level headline
```

This kind of view can be achieved dynamically at display time using `org-indent-mode`, which will prepend intangible space to each line. You can turn on `org-indent-mode` for all files by customizing the variable `org-startup-indented`, or you can turn it on for individual files using

```
#+STARTUP: indent
```

If you want a similar effect in earlier version of Emacs and/or Org, or if you want the indentation to be hard space characters so that the plain text file looks as similar as possible to the Emacs display, Org supports you by helping to indent (with `TAB`) text below each headline, by hiding leading stars, and by only using levels 1, 3, etc to get two characters indentation for each level. To get this support in a file, use

```
#+STARTUP: hidestars odd
```

15.3 MobileOrg

MobileOrg is the name of the mobile companion app for Org mode, currently available for iOS and for Android. *MobileOrg* offers offline viewing and capture support for an Org mode system rooted on a "real" computer. It does also allow you to record changes to existing entries.

The iOS implementation for the *iPhone/iPod Touch/iPad* series of devices, was developed by Richard Moreland. Android users should check out MobileOrg Android by Matt Jones. The two implementations are not identical but offer similar features.

Further reading
Chapter 15 of the manual
Appendix B of the manual
Key reference card

Appendix A GNU Free Documentation License

Version 1.3, 3 November 2008

Copyright © 2000, 2001, 2002, 2007, 2008, 2013, 2014, 2018 Free Software Foundation, Inc.
http://fsf.org/

Everyone is permitted to copy and distribute verbatim copies of this license document, but changing it is not allowed.

0. PREAMBLE

    The purpose of this License is to make a manual, textbook, or other functional and useful document *free* in the sense of freedom: to assure everyone the effective freedom to copy and redistribute it, with or without modifying it, either commercially or noncommercially. Secondarily, this License preserves for the author and publisher a way to get credit for their work, while not being considered responsible for modifications made by others.

    This License is a kind of "copyleft", which means that derivative works of the document must themselves be free in the same sense. It complements the GNU General Public License, which is a copyleft license designed for free software.

    We have designed this License in order to use it for manuals for free software, because free software needs free documentation: a free program should come with manuals providing the same freedoms that the software does. But this License is not limited to software manuals; it can be used for any textual work, regardless of subject matter or whether it is published as a printed book. We recommend this License principally for works whose purpose is instruction or reference.

1. APPLICABILITY AND DEFINITIONS

    This License applies to any manual or other work, in any medium, that contains a notice placed by the copyright holder saying it can be distributed under the terms of this License. Such a notice grants a world-wide, royalty-free license, unlimited in duration, to use that work under the conditions stated herein. The "Document", below, refers to any such manual or work. Any member of the public is a licensee, and is addressed as "you". You accept the license if you copy, modify or distribute the work in a way requiring permission under copyright law.

    A "Modified Version" of the Document means any work containing the Document or a portion of it, either copied verbatim, or with modifications and/or translated into another language.

    A "Secondary Section" is a named appendix or a front-matter section of the Document that deals exclusively with the relationship of the publishers or authors of the Document to the Document's overall subject (or to related matters) and contains nothing that could fall directly within that overall subject. (Thus, if the Document is in part a textbook of mathematics, a Secondary Section may not explain any mathematics.) The relationship could be a matter of historical connection with the subject or with related matters, or of legal, commercial, philosophical, ethical or political position regarding them.

The "Invariant Sections" are certain Secondary Sections whose titles are designated, as being those of Invariant Sections, in the notice that says that the Document is released under this License. If a section does not fit the above definition of Secondary then it is not allowed to be designated as Invariant. The Document may contain zero Invariant Sections. If the Document does not identify any Invariant Sections then there are none.

The "Cover Texts" are certain short passages of text that are listed, as Front-Cover Texts or Back-Cover Texts, in the notice that says that the Document is released under this License. A Front-Cover Text may be at most 5 words, and a Back-Cover Text may be at most 25 words.

A "Transparent" copy of the Document means a machine-readable copy, represented in a format whose specification is available to the general public, that is suitable for revising the document straightforwardly with generic text editors or (for images composed of pixels) generic paint programs or (for drawings) some widely available drawing editor, and that is suitable for input to text formatters or for automatic translation to a variety of formats suitable for input to text formatters. A copy made in an otherwise Transparent file format whose markup, or absence of markup, has been arranged to thwart or discourage subsequent modification by readers is not Transparent. An image format is not Transparent if used for any substantial amount of text. A copy that is not "Transparent" is called "Opaque".

Examples of suitable formats for Transparent copies include plain ASCII without markup, Texinfo input format, LaTeX input format, SGML or XML using a publicly available DTD, and standard-conforming simple HTML, PostScript or PDF designed for human modification. Examples of transparent image formats include PNG, XCF and JPG. Opaque formats include proprietary formats that can be read and edited only by proprietary word processors, SGML or XML for which the DTD and/or processing tools are not generally available, and the machine-generated HTML, PostScript or PDF produced by some word processors for output purposes only.

The "Title Page" means, for a printed book, the title page itself, plus such following pages as are needed to hold, legibly, the material this License requires to appear in the title page. For works in formats which do not have any title page as such, "Title Page" means the text near the most prominent appearance of the work's title, preceding the beginning of the body of the text.

The "publisher" means any person or entity that distributes copies of the Document to the public.

A section "Entitled XYZ" means a named subunit of the Document whose title either is precisely XYZ or contains XYZ in parentheses following text that translates XYZ in another language. (Here XYZ stands for a specific section name mentioned below, such as "Acknowledgements", "Dedications", "Endorsements", or "History".) To "Preserve the Title" of such a section when you modify the Document means that it remains a section "Entitled XYZ" according to this definition.

The Document may include Warranty Disclaimers next to the notice which states that this License applies to the Document. These Warranty Disclaimers are considered to be included by reference in this License, but only as regards disclaiming warranties: any other implication that these Warranty Disclaimers may have is void and has no effect on the meaning of this License.

Appendix A: GNU Free Documentation License

2. VERBATIM COPYING

You may copy and distribute the Document in any medium, either commercially or noncommercially, provided that this License, the copyright notices, and the license notice saying this License applies to the Document are reproduced in all copies, and that you add no other conditions whatsoever to those of this License. You may not use technical measures to obstruct or control the reading or further copying of the copies you make or distribute. However, you may accept compensation in exchange for copies. If you distribute a large enough number of copies you must also follow the conditions in section 3.

You may also lend copies, under the same conditions stated above, and you may publicly display copies.

3. COPYING IN QUANTITY

If you publish printed copies (or copies in media that commonly have printed covers) of the Document, numbering more than 100, and the Document's license notice requires Cover Texts, you must enclose the copies in covers that carry, clearly and legibly, all these Cover Texts: Front-Cover Texts on the front cover, and Back-Cover Texts on the back cover. Both covers must also clearly and legibly identify you as the publisher of these copies. The front cover must present the full title with all words of the title equally prominent and visible. You may add other material on the covers in addition. Copying with changes limited to the covers, as long as they preserve the title of the Document and satisfy these conditions, can be treated as verbatim copying in other respects.

If the required texts for either cover are too voluminous to fit legibly, you should put the first ones listed (as many as fit reasonably) on the actual cover, and continue the rest onto adjacent pages.

If you publish or distribute Opaque copies of the Document numbering more than 100, you must either include a machine-readable Transparent copy along with each Opaque copy, or state in or with each Opaque copy a computer-network location from which the general network-using public has access to download using public-standard network protocols a complete Transparent copy of the Document, free of added material. If you use the latter option, you must take reasonably prudent steps, when you begin distribution of Opaque copies in quantity, to ensure that this Transparent copy will remain thus accessible at the stated location until at least one year after the last time you distribute an Opaque copy (directly or through your agents or retailers) of that edition to the public.

It is requested, but not required, that you contact the authors of the Document well before redistributing any large number of copies, to give them a chance to provide you with an updated version of the Document.

4. MODIFICATIONS

You may copy and distribute a Modified Version of the Document under the conditions of sections 2 and 3 above, provided that you release the Modified Version under precisely this License, with the Modified Version filling the role of the Document, thus licensing distribution and modification of the Modified Version to whoever possesses a copy of it. In addition, you must do these things in the Modified Version:

Appendix A: GNU Free Documentation License

A. Use in the Title Page (and on the covers, if any) a title distinct from that of the Document, and from those of previous versions (which should, if there were any, be listed in the History section of the Document). You may use the same title as a previous version if the original publisher of that version gives permission.
B. List on the Title Page, as authors, one or more persons or entities responsible for authorship of the modifications in the Modified Version, together with at least five of the principal authors of the Document (all of its principal authors, if it has fewer than five), unless they release you from this requirement.
C. State on the Title page the name of the publisher of the Modified Version, as the publisher.
D. Preserve all the copyright notices of the Document.
E. Add an appropriate copyright notice for your modifications adjacent to the other copyright notices.
F. Include, immediately after the copyright notices, a license notice giving the public permission to use the Modified Version under the terms of this License, in the form shown in the Addendum below.
G. Preserve in that license notice the full lists of Invariant Sections and required Cover Texts given in the Document's license notice.
H. Include an unaltered copy of this License.
I. Preserve the section Entitled "History", Preserve its Title, and add to it an item stating at least the title, year, new authors, and publisher of the Modified Version as given on the Title Page. If there is no section Entitled "History" in the Document, create one stating the title, year, authors, and publisher of the Document as given on its Title Page, then add an item describing the Modified Version as stated in the previous sentence.
J. Preserve the network location, if any, given in the Document for public access to a Transparent copy of the Document, and likewise the network locations given in the Document for previous versions it was based on. These may be placed in the "History" section. You may omit a network location for a work that was published at least four years before the Document itself, or if the original publisher of the version it refers to gives permission.
K. For any section Entitled "Acknowledgements" or "Dedications", Preserve the Title of the section, and preserve in the section all the substance and tone of each of the contributor acknowledgements and/or dedications given therein.
L. Preserve all the Invariant Sections of the Document, unaltered in their text and in their titles. Section numbers or the equivalent are not considered part of the section titles.
M. Delete any section Entitled "Endorsements". Such a section may not be included in the Modified Version.
N. Do not retitle any existing section to be Entitled "Endorsements" or to conflict in title with any Invariant Section.
O. Preserve any Warranty Disclaimers.

If the Modified Version includes new front-matter sections or appendices that qualify as Secondary Sections and contain no material copied from the Document, you may at

Appendix A: GNU Free Documentation License

your option designate some or all of these sections as invariant. To do this, add their titles to the list of Invariant Sections in the Modified Version's license notice. These titles must be distinct from any other section titles.

You may add a section Entitled "Endorsements", provided it contains nothing but endorsements of your Modified Version by various parties—for example, statements of peer review or that the text has been approved by an organization as the authoritative definition of a standard.

You may add a passage of up to five words as a Front-Cover Text, and a passage of up to 25 words as a Back-Cover Text, to the end of the list of Cover Texts in the Modified Version. Only one passage of Front-Cover Text and one of Back-Cover Text may be added by (or through arrangements made by) any one entity. If the Document already includes a cover text for the same cover, previously added by you or by arrangement made by the same entity you are acting on behalf of, you may not add another; but you may replace the old one, on explicit permission from the previous publisher that added the old one.

The author(s) and publisher(s) of the Document do not by this License give permission to use their names for publicity for or to assert or imply endorsement of any Modified Version.

5. COMBINING DOCUMENTS

   You may combine the Document with other documents released under this License, under the terms defined in section 4 above for modified versions, provided that you include in the combination all of the Invariant Sections of all of the original documents, unmodified, and list them all as Invariant Sections of your combined work in its license notice, and that you preserve all their Warranty Disclaimers.

   The combined work need only contain one copy of this License, and multiple identical Invariant Sections may be replaced with a single copy. If there are multiple Invariant Sections with the same name but different contents, make the title of each such section unique by adding at the end of it, in parentheses, the name of the original author or publisher of that section if known, or else a unique number. Make the same adjustment to the section titles in the list of Invariant Sections in the license notice of the combined work.

   In the combination, you must combine any sections Entitled "History" in the various original documents, forming one section Entitled "History"; likewise combine any sections Entitled "Acknowledgements", and any sections Entitled "Dedications". You must delete all sections Entitled "Endorsements."

6. COLLECTIONS OF DOCUMENTS

   You may make a collection consisting of the Document and other documents released under this License, and replace the individual copies of this License in the various documents with a single copy that is included in the collection, provided that you follow the rules of this License for verbatim copying of each of the documents in all other respects.

   You may extract a single document from such a collection, and distribute it individually under this License, provided you insert a copy of this License into the extracted document, and follow this License in all other respects regarding verbatim copying of that document.

Appendix A: GNU Free Documentation License

7. AGGREGATION WITH INDEPENDENT WORKS

    A compilation of the Document or its derivatives with other separate and independent documents or works, in or on a volume of a storage or distribution medium, is called an "aggregate" if the copyright resulting from the compilation is not used to limit the legal rights of the compilation's users beyond what the individual works permit. When the Document is included in an aggregate, this License does not apply to the other works in the aggregate which are not themselves derivative works of the Document.

    If the Cover Text requirement of section 3 is applicable to these copies of the Document, then if the Document is less than one half of the entire aggregate, the Document's Cover Texts may be placed on covers that bracket the Document within the aggregate, or the electronic equivalent of covers if the Document is in electronic form. Otherwise they must appear on printed covers that bracket the whole aggregate.

8. TRANSLATION

    Translation is considered a kind of modification, so you may distribute translations of the Document under the terms of section 4. Replacing Invariant Sections with translations requires special permission from their copyright holders, but you may include translations of some or all Invariant Sections in addition to the original versions of these Invariant Sections. You may include a translation of this License, and all the license notices in the Document, and any Warranty Disclaimers, provided that you also include the original English version of this License and the original versions of those notices and disclaimers. In case of a disagreement between the translation and the original version of this License or a notice or disclaimer, the original version will prevail.

    If a section in the Document is Entitled "Acknowledgements", "Dedications", or "History", the requirement (section 4) to Preserve its Title (section 1) will typically require changing the actual title.

9. TERMINATION

    You may not copy, modify, sublicense, or distribute the Document except as expressly provided under this License. Any attempt otherwise to copy, modify, sublicense, or distribute it is void, and will automatically terminate your rights under this License.

    However, if you cease all violation of this License, then your license from a particular copyright holder is reinstated (a) provisionally, unless and until the copyright holder explicitly and finally terminates your license, and (b) permanently, if the copyright holder fails to notify you of the violation by some reasonable means prior to 60 days after the cessation.

    Moreover, your license from a particular copyright holder is reinstated permanently if the copyright holder notifies you of the violation by some reasonable means, this is the first time you have received notice of violation of this License (for any work) from that copyright holder, and you cure the violation prior to 30 days after your receipt of the notice.

    Termination of your rights under this section does not terminate the licenses of parties who have received copies or rights from you under this License. If your rights have been terminated and not permanently reinstated, receipt of a copy of some or all of the same material does not give you any rights to use it.

Appendix A: GNU Free Documentation License

10. FUTURE REVISIONS OF THIS LICENSE

    The Free Software Foundation may publish new, revised versions of the GNU Free Documentation License from time to time. Such new versions will be similar in spirit to the present version, but may differ in detail to address new problems or concerns. See http://www.gnu.org/copyleft/.

    Each version of the License is given a distinguishing version number. If the Document specifies that a particular numbered version of this License "or any later version" applies to it, you have the option of following the terms and conditions either of that specified version or of any later version that has been published (not as a draft) by the Free Software Foundation. If the Document does not specify a version number of this License, you may choose any version ever published (not as a draft) by the Free Software Foundation. If the Document specifies that a proxy can decide which future versions of this License can be used, that proxy's public statement of acceptance of a version permanently authorizes you to choose that version for the Document.

11. RELICENSING

    "Massive Multiauthor Collaboration Site" (or "MMC Site") means any World Wide Web server that publishes copyrightable works and also provides prominent facilities for anybody to edit those works. A public wiki that anybody can edit is an example of such a server. A "Massive Multiauthor Collaboration" (or "MMC") contained in the site means any set of copyrightable works thus published on the MMC site.

    "CC-BY-SA" means the Creative Commons Attribution-Share Alike 3.0 license published by Creative Commons Corporation, a not-for-profit corporation with a principal place of business in San Francisco, California, as well as future copyleft versions of that license published by that same organization.

    "Incorporate" means to publish or republish a Document, in whole or in part, as part of another Document.

    An MMC is "eligible for relicensing" if it is licensed under this License, and if all works that were first published under this License somewhere other than this MMC, and subsequently incorporated in whole or in part into the MMC, (1) had no cover texts or invariant sections, and (2) were thus incorporated prior to November 1, 2008.

    The operator of an MMC Site may republish an MMC contained in the site under CC-BY-SA on the same site at any time before August 1, 2009, provided the MMC is eligible for relicensing.

Appendix A: GNU Free Documentation License

ADDENDUM: How to use this License for your documents

To use this License in a document you have written, include a copy of the License in the document and put the following copyright and license notices just after the title page:

```
Copyright (C) year your name.
Permission is granted to copy, distribute and/or modify this document
under the terms of the GNU Free Documentation License, Version 1.3
or any later version published by the Free Software Foundation;
with no Invariant Sections, no Front-Cover Texts, and no Back-Cover
Texts.  A copy of the license is included in the section entitled ``GNU
Free Documentation License''.
```

If you have Invariant Sections, Front-Cover Texts and Back-Cover Texts, replace the "with...Texts." line with this:

```
with the Invariant Sections being list their titles, with
the Front-Cover Texts being list, and with the Back-Cover Texts
being list.
```

If you have Invariant Sections without Cover Texts, or some other combination of the three, merge those two alternatives to suit the situation.

If your document contains nontrivial examples of program code, we recommend releasing these examples in parallel under your choice of free software license, such as the GNU General Public License, to permit their use in free software.

Lightning Source UK Ltd.
Milton Keynes UK
UKHW020946250422
402015UK00005B/360